SpringerBriefs in Speech Technology

Studies in Speech Signal Processing, Natural Language
Understanding, and Machine Learning

Series Editor:
Amy Neustein
Apt 1809, Fort Lee, NJ, USA

SpringerBriefs present concise summaries of cutting-edge research and practical applications across a wide spectrum of fields. Featuring compact volumes of 50 to 125 pages, the series covers a range of content from professional to academic. Typical topics might include:

- A timely report of state-of-the-art analytical techniques
- A bridge between new research results, as published in journal articles, and a contextual literature review
- A snapshot of a hot or emerging topic
- An in-depth case study or clinical example
- A presentation of core concepts that students must understand in order to make independent contributions

Briefs are characterized by fast, global electronic dissemination, standard publishing contracts, standardized manuscript preparation and formatting guidelines, and expedited production schedules.

The goal of the **SpringerBriefs in Speech Technology** series is to serve as an important reference guide for speech developers, system designers, speech engineers and other professionals in academia, government and the private sector. To accomplish this task, the series will showcase the latest findings in speech technology, ranging from a comparative analysis of contemporary methods of speech parameterization to recent advances in commercial deployment of spoken dialog systems.

More information about this series at http://www.springer.com/series/10043

Pranab Kumar Dhar • Tetsuya Shimamura

Advances in Audio Watermarking Based on Matrix Decomposition

 Springer

Pranab Kumar Dhar
Department of Computer Science and
Engineering
Chittagong University of Engineering and
Technology (CUET)
Chittagong, Bangladesh

Tetsuya Shimamura
Graduate School of Science and
Engineering
Saitama University
Saitama, Japan

ISSN 2191-737X   ISSN 2191-7388   (electronic)
SpringerBriefs in Speech Technology
ISBN 978-3-030-15725-8   ISBN 978-3-030-15726-5   (eBook)
https://doi.org/10.1007/978-3-030-15726-5

Library of Congress Control Number: 2019935575

This Springer imprint is published by the registered company Springer Nature Switzerland AG.
The registered company address is: Gewerbestrasse 11, 6330 Cham, Switzerland

This book is dedicated to my father late Professor Anil Kanti Dhar for his affection, love, encouragement, divine inspiration, and endless support.

Preface

Digital watermarking is identified as a major technique for copyright protection of multimedia contents. It is a process of embedding watermarks into the multimedia data to show authenticity and ownership. This technique has several applications such as copyright protection, data authentication, fingerprinting, data indexing, and broadcast monitoring.

This book introduces audio watermarking methods in transform domain based on matrix decomposition for copyright protection. A blind lifting wavelet transform (LWT)-based watermarking method using fast Walsh-Hadamard transform (FWHT) and singular value decomposition (SVD) for audio copyright protection is proposed. Initially, we preprocess the watermark data to enhance the security of the proposed method. Then, the original audio is segmented into nonoverlapping frames. Watermark data is embedded into the largest singular value of the FWHT coefficients obtained from the low-frequency LWT coefficients of each frame using a quantization function.

A blind audio watermarking method based on LWT and QR decomposition (QRD) for audio copyright protection is introduced. In our proposed method, initially the original audio is segmented into nonoverlapping frames. Watermark information is embedded into the largest element of the upper triangular matrix obtained from the low-frequency LWT coefficients of each frame. A blind watermark detection technique is introduced to identify the embedded watermark under various attacks.

An audio watermarking algorithm based on FWHT and LU decomposition (LUD) is presented. To the best of our knowledge, this is the first audio watermarking method based on FWHT, LUD, and quantization jointly. Initially, we preprocess the watermark data to enhance the security of the proposed algorithm. Then, the original audio is segmented into nonoverlapping frames, and FWHT is applied to each frame. LUD is applied to the FWHT coefficients represented in a matrix form. Watermark data is embedded into the largest element of the upper triangular matrix obtained from the FWHT coefficients of each frame.

An audio watermarking method based on LWT and Schur decomposition (SD) is proposed. To the best of our knowledge, this is the first audio watermarking method based on LWT, SD, and quantization jointly. Initially, the watermark data is preprocessed to enhance the confidentiality of the proposed method. Then, the original audio is segmented into nonoverlapping frames, and LWT is applied to each frame. SD is applied to the selected low-frequency LWT coefficients represented in a matrix form. Watermark data is embedded into the largest element of the upper triangular matrix obtained from the selected LWT coefficients of each frame.

The performance of the proposed watermarking methods is evaluated and, finally, compared with the state-of-the-art methods. Simulation results indicate that the proposed watermarking methods are highly robust against different attacks. In addition, they have high data payload and provide good imperceptible watermarked sounds. Moreover, they outperform the state-of-the-art methods in terms of robustness, imperceptibility, and data payload. These results verify the effectiveness of the proposed methods as a suitable candidate for copyright protection of audio signal.

Keywords Audio watermarking, Copyright protection, Lifting wavelet transform, Fast Walsh-Hadamard transform, Singular value decomposition, QR decomposition, LU decomposition, Schur decomposition

Chittagong, Bangladesh
Saitama, Japan

Pranab Kumar Dhar
Tetsuya Shimamura

Acknowledgments

First of all, I would like to express my sincere gratitude to my Paramaradhya Gurudev Shrimat Maharshi Sutejananda Jyoti Maharaj and Shrimat Swami Sushidhananda Saraswati Maharaj for their endless mercy.

I would also like to express my deep appreciation, obligation, and indebtedness to Prof. Tetsuya Shimamura, my Ph.D. advisor, for his cooperation, encouragement, attention to details, and guidance. I am very grateful to him for his valuable comments and enlightening discussion we had. It has been an honor and a pleasure to work with him during my stay at Saitama University, Japan.

I am extremely thankful to Prof. Jong-Myon Kim, my former advisor. I am also thankful to Prof. Ui-Pil Chong and Prof. Sang-Jin Cho. I am indebted to them for many help, support, and encouragement during my stay at the University of Ulsan, South Korea.

I also extend many thanks to all members of Shimamura Laboratory, both past and present, for their cooperation and friendship.

I am also grateful to all my family members especially my parents late Prof. Anil Kanti Dhar and Mrs. Juthika Dhar for their inspiration, encouragement, and support. I am also thankful to all my friends for their moral support.

Last but not least, I am forever indebted to my wife Uma Dhar, for her love, support, patience, and divine inspiration. I am also thankful to my daughter Parama Dhar who has brightened my life. This book would not have been possible without their unending love and support.

Contents

Chapter 1
Introduction

1.1 Overview

The advent of the Internet revolution and digital multimedia technology has made it extremely convenient for the user to transmit or distribute multimedia (audio, image, and video) data. This has become a serious threat for multimedia content owners. Thus, there is significant interest for copyright protection of multimedia data. Digital watermarking is a process of embedding watermark into the audio data to show authenticity and ownership [1–4]. This technique has several applications such as copyright protection, data authentication, data indexing, broadcast monitoring, and so on.

The basic idea of watermarking is to add a watermark signal into the host data to be watermarked such that the watermark signal is secure in the signal mixture, but can partly or fully be recovered from the signal mixture later on if the correct cryptographically secure key is used. In order to ensure robustness, the watermark information is distributed redundantly over many samples of the host data. Therefore, the recovery is more robust if more watermarked data are available for recovery. In general, watermark systems use one or more cryptographically secure keys to ensure security against manipulation and erasure of the watermark. The embedding process of a generic watermarking scheme is shown in Fig. 1.1. The input to the scheme is the watermark, the host data, and an optional public or secret key. Depending on applications, the host data may be uncompressed or compressed; however, most of the methods work on uncompressed data. The watermark can be of any nature, such as a number, a text, or an image. The secret or public key is used to enforce security. If the watermark is not to be read by unauthorized parties, a key can be used to protect the watermark, in combination with a secret or public key; watermarking techniques are usually referred to as secret and public watermarking techniques, respectively. The output of the watermarking scheme is the watermarked data. The detection process of a generic watermarking scheme is shown in Fig. 1.2.

P. K. Dhar, T. Shimamura, *Advances in Audio Watermarking Based on Matrix Decomposition*, SpringerBriefs in Speech Technology, https://doi.org/10.1007/978-3-030-15726-5_1

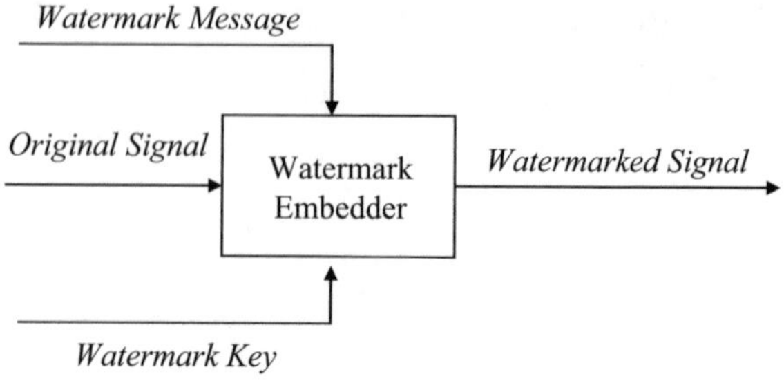

Fig. 1.1 Watermark
embedding process

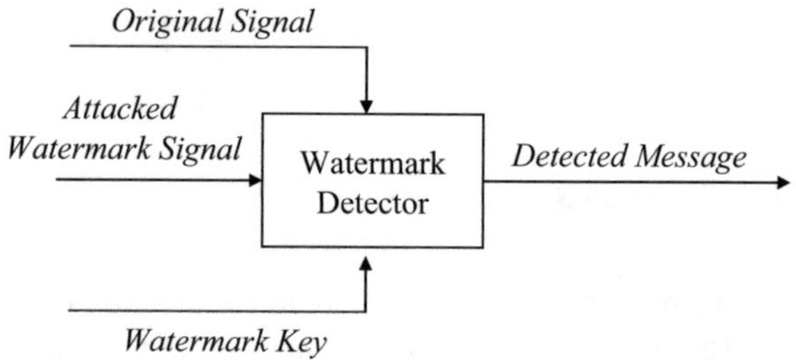

Watermark Message

Original Signal → Watermark Embedder → *Watermarked Signal*

Watermark Key

Fig. 1.2 Watermark
detection process

Original Signal

Attacked Watermark Signal → Watermark Detector → *Detected Message*

Watermark Key

Inputs to the scheme are the attacked watermark data, the secret or public key, and the original data or the original watermark depending on the method. The output of the watermark recovery process is either the recovered watermark or some kind of confidence measure indicating how likely it is for the given watermark at the input to be present in the data under inspection.

1.2   Application Areas of Digital Watermarking

Digital watermarking has various application areas such as copyright protection, fingerprinting, content authentication, copy protection, broadcast monitoring, infor-mation carrier, medical application, and so on. These are discussed in the following.

1.2.1   Copyright Protection

Copyright protection is the most important application of digital watermarking. The objective is to embed information that identifies the copyright owner of the digital media, in order to prevent other parties from claiming the copyright [1, 2, 5, 6].

1.2.2   Fingerprinting

The objective of fingerprinting is to convey information about the legal recipient rather than the source of digital media, in order to identify single distributed copies of digital data [1, 5, 6].

1.2.3   Content Authentication

The objective of content authentication is to detect the modification of digital data. This can be achieved with fragile watermarking technique that has a low robustness to certain modifications [5, 6].

1.2.4   Copy Protection

Copy protection tries to find a mechanism to disallow unauthorized copying of digital media. Copy software or device must be able to detect the watermark and allow or disallow the requested operation according to the copy status of the digital media being copied [1, 5, 6].

1.2.5   Broadcast Monitoring

Watermarking techniques can be useful for broadcast monitoring. Using watermarking, an identification code can be embedded into the content being broadcasted. A computer-based monitoring system can detect the embedded watermark to ensure that they receive all of the airtime they purchase from the broadcasters [1, 5, 6].

1.2.6   Information Carrier

The embedded watermark in this application is expected to have a high capacity and to be detected using a blind detection algorithm. While the robustness against intentional attack is not required, a certain degree of robustness against common attacks may be desired [1, 5, 6].

1.3   Properties of Digital Watermarking

A watermarking method can be characterized by a number of properties. However, the relative importance of each property depends on the demand of the application [1, 5, 6]. These properties are discussed in the following.

1.3.1   Perceptual Transparency

In almost every application, the watermark embedding process has to insert watermark information without changing the perceptual quality of the original signal. The fidelity of a watermarking algorithm is usually defined as a perceptual similarity between the original and watermarked audio signal [1, 5, 6].

1.3.2   Data Payload

The data payload of a watermarking method is the number of watermark bits that are embedded within a unit of time and is usually measured in bits per second (bps). Some watermarking applications, such as copy control, require the insertion of a serial number or an author ID, with the average bit rate of 0.5 bps [1, 5, 6].

1.3.3   Robustness

The robustness of a watermarking method is defined as the ability to detect the watermark after common signal processing operations. The requirement of robustness of a watermarking method is completely application dependent [1, 5, 6].

1.3.4   Blind or Informed Detection

In some applications, the detection algorithm can use the original audio signal to extract watermark from the watermarked signal (informed detection). However, if the detection algorithm does not have access to the original signal (blind detection) and this inability substantially decreases the amount of data that can be hidden in the original signal [1, 5, 6].

1.3.5   Security

The security of digital watermarking is interpreted in the same way as the security of encryption techniques and it cannot be broken unless the authorized user has access to a secret key that controls watermark embedding. An unauthorized user should be unable to extract the data in a reasonable amount of time even if he or she knows that the original signal contains a watermark and is familiar with the watermark embedding algorithm [1, 5, 6].

1.3.6   Computational Complexity

The principal issues from the technical point of view are the computational complexity of embedding and detection algorithms and the number of embedders and detectors used in the system. One of the economic issues in the design of embedders and detectors, which can be implemented either as hardware or software plugins, is the difference in processing power of different devices (laptop, mobile phone, etc.) [1, 5, 6].

1.4   Related Research

A comprehensive survey on audio watermarking can be found in [1–5]. Most audio watermarking methods can be categorized into time domain methods [7, 8] or the transform domain methods such as discrete wavelet transform (DWT) [9–11], discrete cosine transform (DCT) [12, 13], discrete sine transform (DST) [14], lifting wavelet transform (LWT) [15], and fast Fourier transform (FFT) [16, 17]. Time domain methods are very efficient and easy to implement, although transform domain methods can provide high robustness. Lie and Chang [7] introduced a method in which group amplitudes are modified to achieve high robustness. Bassia et al. [8] presented a watermarking technique in which watermark bits are embedded by modifying the audio samples directly. However, both methods have low data payload. In [9], authors presented an adaptive method using wavelet-based entropy, but robustness to re-sampling and low-pass filtering attacks are quite low. Chen et al. [10] proposed an algorithm that embeds watermark information by energy-proportion scheme. However, the signal-to-noise ratio (SNR) results of this algorithm are not satisfactory. In [11], authors introduced an optimization-based watermarking scheme that embeds watermark in the lowest-frequency coefficients of DWT. However, the subjective evaluation of watermarked audio signals has not been conducted in this scheme. Erçelebi and Batakçı [14] proposed a watermarking method based on LWT in which a binary image is embedded as watermark. However, from the reported result, robustness to attacks of this method is quite low. Megías et al. [15] suggested a watermarking method that embeds watermark in FFT domain, but it has low bit error rate (BER) against some attacks. Some novel and popular audio watermarking methods use the patchwork algorithm [18–20]. Recently, the singular value decomposition (SVD) has been used as an effective technique in digital watermarking [21–25]. Dhar et al. [21] proposed an efficient audio watermarking algorithm in transform domain based on SVD and Cartesian-polar transform (CPT). However, the detection scheme is non-blind and robustness needs further improvement. In [22], authors proposed a blind SVD-based method using entropy and log-polar transform (LPT). However, robustness against re-sampling attack is little low. The methods proposed by Lei et al. [23] and Bhat et al. [24] provide high robustness against attack; however, the data payload of these methods is quite low. Al-Nuaimy et al. [25] proposed an efficient SVD-based audio watermarking scheme in the transform domain that utilizes a chaotic sequence to shuffle the binary watermark to increase the confidentiality. Moreover, they extended the proposed watermarking method and applied it in Bluetooth-based systems and automatic speaker identification systems. However, from the reported results, the robustness needs further improvement. Moreover, some other techniques such as empirical mode decomposition (EMD) [26], time spread (TS) echo method [27], and audio histogram [28, 29] techniques are becoming popular in audio watermarking field. The major limitation of the existing audio watermarking techniques is the difficulty in obtaining a favorable trade-off between the data payload and robustness against various attacks while maintaining the perceptual quality of the watermarked audio signal at an acceptable level. To overcome this limitation, in this book, the following audio watermarking methods are proposed:

- A blind lifting wavelet transform (LWT) based watermarking method using fast Walsh-Hadamard transform (FWHT) and singular value decomposition (SVD) for audio copyright protection is proposed. The main features of the proposed method are: (*i*) it utilizes the LWT, FWHT, and SVD jointly; (*ii*) it uses Bernoulli map, containing the chaotic characteristic to enhance the confidentiality of the proposed method; (*iii*) watermark data are embedded into the largest singular value of the FWHT coefficients obtained from the low-frequency LWT coefficients of each frame using a quantization function; (*iv*) watermark extraction process is blind; (*v*) subjective and objective evaluations reveal that the proposed method maintains high audio quality; and (*vi*) it achieves a good trade-off among imperceptibility, robustness, and data payload.
- A blind audio watermarking method based on LWT and QR decomposition (QRD) for audio copyright protection is introduced. In our proposed method, initially the original audio is segmented into nonoverlapping frames. Watermark information is embedded into the largest element of the upper triangular matrix obtained from the low-frequency LWT coefficients of each frame. A blind watermark detection technique is introduced to identify the embedded watermark under various attacks.

- An audio watermarking algorithm based on FWHT and LU decomposition (LUD) is presented. To the best of our knowledge, this is the first audio watermarking method based on FWHT, LUD, and quantization jointly. Initially, we preprocess the watermark data to enhance the security of the proposed algorithm. Then, the original audio is segmented into nonoverlapping frames and FWHT is applied to each frame. LUD is applied to the FWHT coefficients represented in a matrix form. Watermark data are embedded into the largest element of the upper triangular matrix obtained from the FWHT coefficients of each frame.
- An audio watermarking method based on LWT and Schur decomposition (SD) is proposed. To the best of our knowledge, this is the first audio watermarking method based on LWT, SD, and quantization jointly. Initially, the watermark data is preprocessed to enhance the confidentiality of the proposed method. Then, the original audio is segmented into nonoverlapping frames and LWT is applied to each frame. SD is applied to the selected low-frequency LWT coefficients represented in a matrix form. Watermark data are embedded into the largest element of the upper triangular matrix obtained from the selected LWT coefficients of each frame.

1.5   Signal Transformation

In this section, some signal transformation techniques used in the proposed methods are discussed briefly.

1.5.1   Lifting Wavelet Transform

The LWT is designed to reduce the computation time and memory requirement. It has several unique properties in comparison with traditional wavelet: (*i*) it allows an in-place implementation of the fast wavelet transform and thus it can be calculated more efficiently and needs less memory space, (*ii*) it is particularly easy to build nonlinear wavelet transforms, (*iii*) it has the frequency localization features that overcome the weakness of the traditional wavelet transform. The main principle of the lifting wavelet is to construct a new wavelet with better characteristics based on a simple wavelet. Lifting wavelet scheme consists of three steps: split/merge, prediction, and update. The detailed reasoning and proof of the lifting scheme is given in [9].

1.5.2 Fast Walsh-Hadamard Transform

The FWHT is widely used in various applications, such as image processing, signal processing, filtering, and so on. The forward and inverse FWHT can be defined as a linear combination of a set of square waves of different frequencies and are represented by the following equations:

$$X_w(k) = \sum_{n=0}^{N-1} x(n) w_N(k,n)$$
$$= \sum_{n=0}^{N-1} x(n) \prod_{i=0}^{M-1} (-1)^{n_i k_{M-1-i}}, k = 0,1,.....N-1 \tag{1.1}$$

$$x(n) = \frac{1}{N} \sum_{n=0}^{N-1} X_w(k) w_N(k,n)$$
$$= \frac{1}{N} \sum_{n=0}^{N-1} X_w(k) \prod_{i=0}^{M-1} (-1)^{n_i k_{M-1-i}}, n = 0,1,.....N-1 \tag{1.2}$$

where $x(n)$ is the input signal in time domain, $X_w(k)$ is the transformed signal, $w_N(k,n)$ is the walsh function, $N = 2^n$, $M = \log_2 N$, and n_i is the ith bit in the binary representation of n.

1.6 Matrix Decomposition

1.6.1 Singular Value Decomposition

SVD is a mathematical tool that is mainly used to analyze matrices. In SVD transformation, a given matrix A is decomposed into three matrices. Let $A = \{A_{ij}\}_{p \times p}$ be an arbitrary matrix with SVD of the form $A = USV^T$ where U and V are orthogonal $p \times p$ matrices and S is a $p \times p$ diagonal matrix with nonnegative elements. The diagonal entries of S are called the singular values (SVs) of A where $S = \text{diag}(\sigma_1, \sigma_2, \ldots, \sigma_p)$, the columns of U are called the left singular vectors of A, and the columns of V are called the right singular vectors of A. The SVD has some interesting properties: (*i*) the sizes of the matrices from SVD transformation are not fixed, and the matrices need not be square; (*ii*) changing SVs slightly does not affect the quality of the signal much; (*iii*) the SVs are invariant under common signal processing operations; and (*iv*) the SVs satisfy intrinsic algebraic properties.

1.6.2   QR Decomposition

In QRD, the orthogonal-triangular decomposition of a matrix is performed with the following definition:

$$[Q,R] = qr(A) \tag{1.3}$$

where R is an $m \times m$ upper triangular matrix and Q is an $m \times m$ unitary matrix, so that $A = QR$. The columns of Q are obtained from the columns of A by Gram-Schmidt orthogonalization process. One of the good properties of R matrix is that when columns of A have correlation with each other, the absolute values of the elements in the first row of R matrix are probably greater than those in other rows. This property can be utilized for embedding watermark in audio signal.

1.6.3   Schur Decomposition

SD is a mathematical tool that is mainly used to analyze matrices. It can be defined as follows:

$$[U,S] = Schur(A) \tag{1.4}$$

where A is an $m \times m$ square matrix, S is an $m \times m$ upper triangular matrix, and U is an $m \times m$ unitary matrix, so that $A = USU^{-1}$. Here, U is called a Schur form of A. Since U is similar to A, it has the same multiset of eigenvalues, and since it is triangular, those eigenvalues are the diagonal entries of U.

1.7   Book Organization

This book is divided into six chapters. This chapter briefly discussed a general watermarking scheme, its properties and application areas. In addition, the related research, signal transformation, and matrix decomposition are presented in this chapter. The rest of this book is organized as follows. Chapter 2 introduces a blind lifting wavelet transform (LWT) based watermarking method using fast Walsh-Hadamard transform (FWHT) and singular value decomposition (SVD). Chapter 3 proposes a blind audio watermarking method based on LWT and QR decomposition (QRD). Chapter 4 presents an audio watermarking algorithm based on FWHT and LU decomposition (LUD). Chapter 5 introduces an audio watermarking method based on LWT and Schur decomposition (SD). Chapter 6 concludes the book with brief summary of the key points. A future research direction is also provided in this chapter.

Chapter 2
LWT-Based Audio Watermarking Using FWHT and SVD

2.1 Introduction

This chapter introduces an LWT-based audio watermarking scheme using fast Walsh-Hadamard transform (FWHT) and singular value decomposition (SVD) [30]. Conventional wavelet transform provides good results for its multi-resolution characteristics and perfect reconstruction. However, it is mainly calculated by convolution operation, resulting in high computation. In addition, the generated floating numbers increase the storage requirements. As a result, the LWT is designed to increase the efficiency and it is now used in digital watermarking [27]. In our proposed method, watermark information is preprocessed first using a Bernoulli map in order to improve the robustness and enhance the confidentiality of the watermark. Then the original audio is segmented into nonoverlapping frames. Watermark information is embedded into the largest singular value of the FWHT coefficients obtained from the low-frequency LWT coefficients of each frame. A blind watermark detection technique is developed to identify the embedded watermark under various attacks. The main features of the proposed scheme are: (*i*) it utilizes the LWT, FWHT, and SVD jointly; (*ii*) it uses Bernoulli map, containing the chaotic characteristic to enhance the confidentiality of the proposed scheme; (*iii*) watermark extraction process is blind; (*iv*) subjective and objective evaluations reveal that the proposed scheme maintains high audio quality; and (*v*) it achieves a good trade-off among imperceptibility, robustness, and data payload. Experimental results indicate that the proposed watermarking scheme is highly robust against various attacks such as noise addition, cropping, re-sampling, re-quantization, and MP3 compression. Moreover, it outperforms state-of-the-art methods [9–10, 14–16, 20, 23, 26, 28] in terms of imperceptibility, robustness, and data payload. The data payload of the proposed scheme is 172.39 bps, which is relatively higher than that of the state-of-the-art methods.

© The Author(s), under exclusive license to Springer Nature Switzerland AG 2019
P. K. Dhar, T. Shimamura, *Advances in Audio Watermarking Based on Matrix Decomposition*, SpringerBriefs in Speech Technology,
https://doi.org/10.1007/978-3-030-15726-5_2

2.2   Proposed Watermarking Method

Let $X = \{x(m), 1 \leq m \leq L\}$ be an original audio signal with L samples, $W = \{w(k, l), 1 \leq k \leq P, 1 \leq l \leq P\}$ be a binary logo image to be embedded into the original audio signal.

2.2.1   Watermark Preprocessing

This paper uses a Bernoulli map that contains the chaotic characteristics to encrypt the binary watermark image for enhancing the confidentiality of the proposed scheme. Bernoulli map can be defined as follows:

$$y(i+1) = \begin{cases} 2y(i), 0 \leq y(i) < \frac{1}{2} \\ 2y(i)-1, \frac{1}{2} \leq y(i) \leq 1 \end{cases} \tag{2.1}$$

where $y(1) \in (0,1)$ is a real parameter (map's initial condition). Then $z(i)$ is calculated by using the following equation:

$$z(i) = \text{round}\big(y(i)\big) \tag{2.2}$$

The binary watermark image W is converted into a one-dimensional vector q, where $q = \{q(i), i = 1, 2, 3, \ldots, P \times P\}$. The vector q is then permuted by a specific order to eliminate the spatial correlation of the image pixels and a new sequence $e = \{e(i), i = 1, 2, 3, \ldots, P \times P\}$ is obtained. This specific order is used as secret key C. Finally, $e(i)$ is encrypted by $z(i)$ using the following rule:

$$u(i) = z(i) \oplus e(i), 1 \leq i \leq P \times P \tag{2.3}$$

where $\oplus$ is the exclusive-or (XOR) operation.

Watermark Embedding Process

The proposed watermark embedding process is shown in Fig. 2.1. The embedding process is described as follows:

1. The original audio signal X is first segmented into nonoverlapping frames $F = \{F_1, F_2, F_3, \ldots, F_{P \times P}\}$.
2. A two-level LWT is performed on each frame F_i. This operation produces three sets of coefficients D_1, D_2, and A_2, where D_i and A_i represent the detailed and approximate coefficients, respectively.

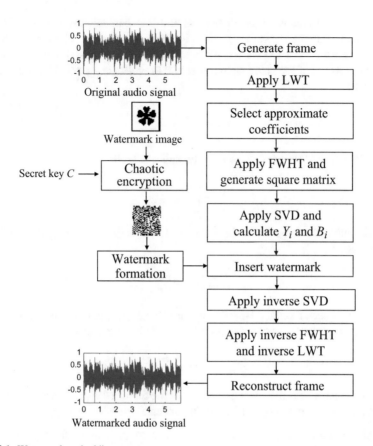

Fig. 2.1 Watermark embedding process

3. The FWHT is applied to the approximate coefficients A_2 of each frame F_i to obtain the FWHT coefficients.
4. The FWHT coefficients of each frame F_i are rearranged into a $K \times K$ square matrix R_i. This is done by dividing the coefficient set into K segments with K coefficients.
5. SVD is performed to decompose each matrix R_i into three matrices: U_i, S_i, and V_i. The SVD operation is represented as follows:

$$R_i = U_i S_i V_i^T \tag{2.4}$$

6. Multiply the highest singular value $S_i(1,1)$ of each frame F_i by a scaling factor β and then round it to the nearest integer. Let $Y_i = \text{round}\ (\beta * S\ (1,\ 1))$ and $B_i = \text{mod}\ (Y_i,\ 2)$. A small value of β will lead to good imperceptibility but will provide low robustness to the attacks. Thus, an optimal value of β will be chosen for trade-off between imperceptibility and robustness.

7. In order to guarantee the robustness and transparency, the proposed scheme embeds watermark bit into Y_i of each matrix S_i using the following equations. This ensures that the watermark is located at the most significant perceptual components of the audio signal. If $B_i = 0$ the following embedding equation is used:

$$Y_i' = \begin{cases} Y_i+1 & \text{if } u(i)=1 \\ Y_i & \text{if } u(i)=0 \end{cases} \tag{2.5}$$

If $B_i = 1$, the following embedding equation is used:

$$Y_i' = \begin{cases} Y_i+1 & \text{if } u(i)=0 \\ Y_i & \text{if } u(i)=1 \end{cases} \tag{2.6}$$

8. The modified largest singular value $S_i'(1,1)$ is calculated using the following equation:

$$S_i'(1,1) = \frac{Y_i'}{\beta} \tag{2.7}$$

9. Reinsert each modified largest singular value $S_i'(1,1)$ into matrix S_i and inverse SVD is applied to obtain the modified matrix R_i' that is given by

$$R_i' = U_i S_i' V_i^T \tag{2.8}$$

Each matrix R_i' is then reshaped to create the modified approximate coefficients A_2' of each frame F_i by performing the inverse operation of step 4.

10. After substituting the modified approximate coefficients A_2' for A_2, a two-level inverse LWT is performed to obtain the watermarked audio frame F_i'.
11. Finally, all watermarked frames are concatenated to calculate the watermarked audio signal X'.

Watermark Detection Process

The proposed watermark detection process is shown in Fig. 2.2 and it does not need the original audio signal to extract the watermark. The detection process is described as follows:

1. A two-level LWT is performed on each frame F_i^* of the attacked watermarked audio signal.
2. The FWHT is applied to the approximate coefficients A_2^* of each frame F_i to obtain the FWHT coefficients. Rearrange each A_2^* to obtain R_i^*.
3. SVD is performed on each R_i^* to get U_i^*, S_i^*, and V_i^*. The largest value $S_i^*(1, 1)$ of each matrix S_i^* of the attacked watermarked audio frame is selected.
4. Calculate Y_i^* and B_i^* of each highest singular value $S_i^*(1, 1)$.

Fig. 2.2 Watermark
detection process

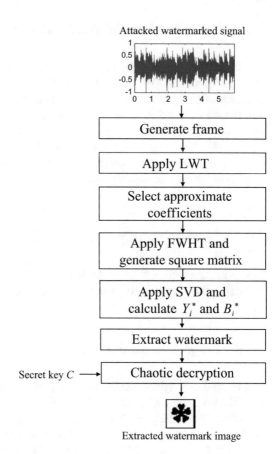

Attacked watermarked signal

Generate frame

Apply LWT

Select approximate
coefficients

Apply FWHT and
generate square matrix

Apply SVD and
calculate Y_i^* and B_i^*

Extract watermark

Secret key C ⟶ Chaotic decryption

Extracted watermark image

5. Watermark sequence is extracted as follows:

$$u^*\left(i\right) = \begin{cases} 1 & \text{if } B_i^*=1 \\ 0 & \text{if } B_i^*=0 \end{cases} \tag{2.9}$$

6. Perform chaotic decryption to find the binary sequence $e^*(i)$ using the following rule:

$$e^*\left(i\right) = z\left(i\right) \oplus u^*\left(i\right) \tag{2.10}$$

7. $e^*(i)$ is then permuted using the secret key C to obtain the hidden binary sequence $q^*(i)$.

8. Finally, watermark image is obtained by rearranging the binary sequence $q^*(i)$ into a square matrix W^* of size $P \times P$.

2.3   Simulation Results and Discussion

In this section, we conducted several experiments on four different types of 16-bit mono audio signals (Pop, Folk, Classical, and Speech) sampled at 44.1 kHz. Each audio file contains 262,144 samples (duration 5.94 s). Each audio signal is divided into frames of size 256 samples. In each frame of audio signal, we embed one bit watermark data of a binary logo image. The binary logo image and the corresponding encrypted image of size $P \times P = 32 \times 32 = 1024$ are shown in Fig. 2.3.

The performance of the proposed scheme was evaluated in terms of imperceptibility, robustness, and data payload.

2.3.1   Imperceptibility Test

The imperceptibility of the watermarked audio signal is evaluated by using two ways: (*i*) subjective listening test (*ii*) objective test.

2.3.2   Subjective Listening Test

An informal subjective listening test was performed to evaluate the perceptual quality of the watermarked audio. Ten listeners involved in this listening test were asked to classify the differences between them, using a five-point subjective difference grade (SDG) ranging from 5.0 to 1.0 (imperceptible to very annoying) as shown in Table 2.1. The average SDG (i.e., mean opinion) scores for different watermarked sounds using the proposed method are shown in Table 2.2. From the test results, we observed that the average mean opinion score (MOS) of all watermarked sound using the proposed method is 4.83, indicating that original and watermarked audio signals are perceptually similar.

Fig. 2.3 (**a**) Binary
watermark image (**b**)
Encrypted watermark
image

a) Binary watermark image (b) Encrypted watermark image

Table 2.1 Subjective and objective difference grades

SDG	ODG	Description	Quality
5	0	Imperceptible	Excellent
4	−1	Perceptible, but not annoying	Good
3	−2	Slightly annoying	Fair
2	−3	Annoying	Poor
1	−4	Very annoying	Bad

Table 2.2 MOS, ODG, and SNR results for different watermarked sounds

Types of Signal	MOS	ODG	SNR
Pop	4.80	−1.08	34.50
Folk	4.90	−0.25	33.73
Classical	4.90	−0.28	33.13
Speech	4.70	−1.23	33.97
Average	*4.83*	*−0.72*	*33.83*

2.3.3   Objective Test

Objective test was conducted using the objective difference grade (ODG), which is one of the output values obtained from the perceptual evaluation of audio quality (PEAQ) measurement technique specified in ITU-R BS.1387 (International Telecommunication Union-Radio-communication Sector) standard [32]. It corresponds to the subjective grade used in human-based audio tests. The ODG ranges from 0.0 to −4.0 (imperceptible to very annoying) as shown in Table 2.1. The objective quality of the watermarked audio signals is calculated in terms of ODG and shown in Table 2.2. We observed that the average ODG value is −0.72, indicating that original and watermarked audio signals are perceptually indistinguishable.

Objective evaluation was also done by calculating the SNR, which is given by

$$\text{SNR} = 10\log_{10}\frac{\sum_{i=1}^{L}x^2(n)}{\sum_{i=1}^{L}\left[x(n)-x^*(n)\right]^2} \tag{2.11}$$

where $x(n)$ and $x^*(n)$ are the original and watermarked audio signals in time domain, respectively. According to the International Federation of the Phonographic Industry (IFPI) standard [13], audio watermarking should be imperceptible when SNR is over 20 dB. After embedding watermark information, the SNRs of the watermarked audio signals using the proposed method are above 20 dB, shown in Table 2.2, which satisfied the IFPI standard. We observed that the MOS, ODG, and SNR values range from 4.7 to 4.9, −1.23 to −0.25, and 33.13 to 34.50, respectively, indicating that the proposed watermarking scheme provides good imperceptible watermarked sound. Table 2.3 shows a comparison between the proposed scheme and the several recent methods in terms of SNR and MOS that are based on the

Table 2.3 SNR and MOS comparison between the proposed scheme and several recent methods

Reference	Algorithm	SNR	MOS
[10]	DWT-based energy proportion	17.95	4.15
[9]	Wavelet-based entropy	22.46	4.38
[26]	EMD	24.12	–
[20]	Patchwork	24.95	4.67
[16]	FFT amplitude modification	25.70	–
[23]	DCT-SVD	32.53	4.71
Proposed	*LWT-FWHT-SVD*	*33.83*	*4.83*

reported results in the references [9–10, 16, 20, 23, 26]. From these results, we observed that our proposed scheme provides better result than the recent watermarking methods in terms of SNR and MOS. In other words, subjective and objective evaluations prove a high transparency of the proposed scheme.

2.3.4 Robustness Test

Various signal processing attacks were performed to assess the robustness of the proposed method.

Normalized correlation (NC) coefficient is used to compare the similarities between the original watermark W and the extracted watermark W^*, which is calculated as

$$\mathrm{NC}\left(W, W^*\right) = \frac{\sum_{i=1}^{I} w(i) \cdot w^*(i)}{\sqrt{\sum_{i=1}^{I} w(i) \cdot w(i)} \sqrt{\sum_{i=1}^{I} w^*(i) \cdot w^*(i)}} \tag{2.12}$$

The correlation between W and W^* is very high when NC (W, W^*) is close to 1. On the other hand, the correlation between W and W^* is very low when NC (W, W^*) is close to zero.

The bit error rate (BER) is used to measure the robustness of a watermarking method and is computed as

$$\mathrm{BER}\left(W, W^*\right) = \frac{\sum_{i=1}^{I} w(i) \oplus w^*(i)}{I} \tag{2.13}$$

where $\oplus$ is an exclusive or (XOR) operation.

Various signal processing attacks summarized in Table 2.4 were performed to assess the robustness of the proposed scheme. Table 2.5 shows the robustness results of the proposed scheme in terms of NC and BER against several attacks for the audio signal "Classical." The NC values are all above 0.96 and the BER values are below 5%. The extracted watermark images are visually similar to the original

Table 2.4 Attacks used in this study for watermarked sound

Attacks	Description
Noise addition	Additive white Gaussian noise (AWGN) is added to the watermarked audio signal.
Cropping	1000 samples are removed from the beginning of the watermarked signal and then these samples are replaced by the watermarked samples attacked with AWGN.
Re-sampling	The watermarked signal originally sampled at 44.1 kHz is re-sampled at 22.050 kHz and then restored by sampling again at 44.1 kHz.
Re-quantization	The 16-bit watermarked audio signal is quantized down to 8 bits/sample and again re-quantized back to 16 bits/sample.
MP3 Compression	MPEG-1 layer 3 compression with 128 kbps is applied to the watermarked audio signal.

Table 2.5 NC and BER of extracted watermark image for the audio signal "Classical"

Attack type	NC	BER (%)	Extracted watermark
No attack	1	0	
Noise addition	0.9951	0.5859	
Cropping	0.9992	0.0977	
Re-sampling	0.9902	1.1719	
Re-quantization	1	0	
MP3 compression	0.9666	4.0039	

watermark. This clearly indicates a good performance of the proposed scheme against different attacks. Table 2.6 shows similar results for the audio signal "Pop," "Folk," and "Speech," respectively. The NC values are all above 0.97 and the BER values are all below 4%, demonstrating the high robustness of our proposed scheme against different attacks.

The proposed scheme utilizes chaotic encryption to enhance the security. The proposed watermark embedding and detection processes depend on the secret key C. Thus, it is impossible to maliciously detect the watermark without this key.

Table 2.7 shows a general comparison between the proposed scheme and several recent methods [10, 14–15, 23, 26, 28] for data payload and robustness to re-sampling, re-quantization and MP3 compression attack. The data payload of the proposed scheme is 172.39 bps. From this comparison, we observed that the proposed scheme provides higher data payload and lower BER values against various attacks compared with the state-of-the-art watermarking methods. This is because watermark bits are embedded into the largest singular value of the FWHT coefficients obtained from the LWT coefficients of each frame.

Table 2.6 NC and BER of the extracted watermark for different audio signals

Audio signal	Attack type	NC	BER (%)
Pop	No attack	1	0
	Noise addition	0.9943	0.6836
	Cropping	0.9984	0.1953
	Re-sampling	0.9910	1.0742
	Re-quantization	1	0
	MP3 compression	0.9737	3.1250
Folk	No attack	1	0
	Noise addition	0.9951	0.5859
	Cropping	0.9976	0.2930
	Re-sampling	0.9894	1.2695
	Re-quantization	1	0
	MP3 compression	0.9721	3.3203
Speech	No attack	1	0
	Noise addition	0.9926	0.8789
	Cropping	0.9984	0.1953
	Re-sampling	0.9918	0.9766
	Re-quantization	1	0
	MP3 Compression	0.9812	2.2461

Table 2.7 A General Comparison Between the Proposed Scheme and Several Recent Methods Sorted by Data Payload

Reference	Algorithm	Payload (bps)	Re-sampling BER (%)	Re-quantization BER (%)	MP3 compression BER (%)
Proposed	*LWT-FWHT-SVD*	*172.39*	*1.27 (22.05 kHz)*	*0 (8 bits/sample)*	*4.01 (128 kbps)*
[10]	Wavelet-based entropy	172.28–86.14	9.1 (22.05 kHz)	–	6.7 (128 kbps)
[14]	Chaos based DFRST	86	0 (22.05 kHz)	–	3.47 (48 kbps)
[26]	EMD	46.9–50.3	3 (22.05 kHz)	0 (8 bits/sample)	1 (32 kbps)
[23]	DCT-SVD	43	0 (22.05 kHz)	0 (8 bits/sample)	3 (32 kbps)
[28]	Histogram	3	0 (–)	0 (8 bits/sample)	15 (128 kbps)
[15]	LWT	–	16.50 (36.750 kHz)	22.09 (8 bits/sample)	51.73 (128 kbps)

The false positive error (FPE) is the probability that an unwatermarked audio signal is declared as watermarked signal by the detector. Then, FPE probability P_{f_p} can be calculated as

$$P_{f_p} = 2^{-k} \sum_{m=\lceil 0.8k \rceil}^{k} \binom{k}{m} \tag{2.14}$$

where $\binom{k}{m}$ is the binomial coefficient, k is the total number of watermark bits and m is the total number of matching bits. Figure 2.4 plots the FPE probability for $k \in (0,100]$. It is noted that P_{f_p} approaches 0 when k is larger than 30.

The false negative error (FNE) is the probability that a watermarked audio signal is declared as unwatermarked signal by the detector. Then FNE P_{f_n} can be calculated as

$$P_{f_n} = \sum_{m=0}^{\lceil 0.8k \rceil - 1} \left[\binom{k}{m} (P)^m (1-P)^{k-m} \right] \tag{2.15}$$

where P is the bit error rate probability of the extracted watermark. The approximate value of P can be obtained from BER under different attacks. Figure 2.5 plots the FNE probability for $k \in (0,100]$. It is noted that P_{f_n} approaches 0 when k is larger than 30.

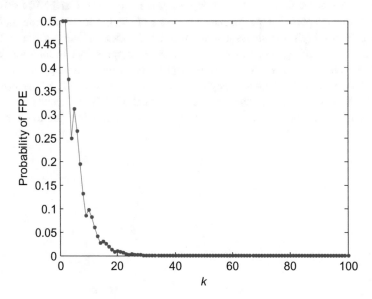

Fig. 2.4 Probability of FPE for various values of k

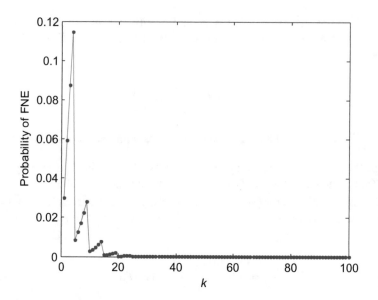

Fig. 2.5 Probability of FNE for various values of k

2.4   Concluding Remarks

A blind LWT-based audio watermarking scheme using FWHT and SVD was proposed in this chapter. The performance analysis of the proposed scheme indicates
that it provides high robustness against different attacks including noise addition,
cropping, re-sampling, re-quantization, and MP3 compression. In addition, audio
quality evaluation tests show high imperceptibility of the watermark in the audio
signal. Moreover, it has high data payload and provides more excellent performance
than the state-of-the-art audio watermarking methods. These results verify the
effectiveness of the proposed watermarking scheme for audio copyright
protection.

Chapter 3
Audio Watermarking Based on LWT and QRD

3.1 Introduction

This chapter presents a blind audio watermarking method based on LWT and QR decomposition (QRD) for audio copyright protection [31]. In our proposed method, initially the original audio is segmented into nonoverlapping frames. Watermark information is embedded into the largest element of the upper triangular matrix obtained from the low-frequency LWT coefficients of each frame. A blind watermark detection technique is introduced to identify the embedded watermark under various attacks. Simulation results indicate that the proposed watermarking method is highly robust against different attacks. In addition, it has high data payload and provides good imperceptible watermarked sounds. Moreover, it shows better result than the state-of-the-art methods in terms of imperceptibility and robustness. In this chapter, a blind audio watermarking method based on LWT and QR decomposition (QRD) is proposed. The main features of the proposed method are: (*i*) it utilizes the LWT and QRD jointly; (*ii*) it uses Bernoulli map, containing the chaotic characteristic to enhance the confidentiality of the proposed method; (*iii*) watermark extraction process is blind; (*iv*) subjective and objective evaluations reveal that the proposed method maintains a high audio quality; and (*v*) it achieves a good trade-off among imperceptibility, robustness, and data payload. Simulation results demonstrate that the proposed watermarking method shows high robustness against various attacks such as noise addition, cropping, re-sampling, re-quantization, and MP3 compression. Moreover, it outperforms the state-of-the-art methods [9–10, 15–16, 20, 23, 24, 26, 29] in terms of imperceptibility, robustness, and data payload. The data payload of the proposed method is 172.39 bps, which is relatively higher than that of the state-of-the-art methods.

© The Author(s), under exclusive license to Springer Nature Switzerland AG 2019   23
P. K. Dhar, T. Shimamura, *Advances in Audio Watermarking Based on Matrix Decomposition*, SpringerBriefs in Speech Technology,
https://doi.org/10.1007/978-3-030-15726-5_3

3.2   Proposed Watermarking Method

Let $X = \{x(n), 1 \leq n \leq L\}$ be an original audio signal with L samples, $W = \{w(i), 1 \leq i \leq I\}$ be a binary watermark sequence to be embedded into the original audio signal.

3.2.1   Watermark Preprocessing

To enhance the security of the proposed method, watermark sequence should be preprocessed first. The proposed method utilizes a Gaussian map, containing the chaotic characteristics to encrypt the binary watermark sequence for enhancing the confidentiality of the proposed method. It can be defined as follows:

$$y(i+1) = \exp\left(-a\left(y(i)\right)^2\right) + b \tag{3.1}$$

where $y(1) \in (0,1)$, a and b are real parameters (map's initial condition). Then $y(i)$ is mapped using the following rule:

$$z(i) = \begin{cases} 1 & \text{if } y(i) > T \\ 0 & \text{otherwise} \end{cases} \tag{3.2}$$

where T is a predefined threshold. Finally, $w(i)$ is encrypted by $z(i)$ with the following rule:

$$u(i) = z(i) \oplus w(i), 1 \leq i \leq I \tag{3.3}$$

where $\oplus$ is the exclusive-or (XOR) operation and $u(i)$ is the encrypted watermark sequence. After applying this chaotic encryption technique, the original watermark is encrypted and cannot be found by random search. In this study, the value of $y(1)$, a, and b are used as secret key K.

3.2.2   Watermark Embedding Process

The proposed watermark embedding process is given in Fig. 3.1 and described as follows:

1. The original audio signal X is first segmented into nonoverlapping frames $F = \{F_1, F_2, F_3, \ldots, F_I\}$.

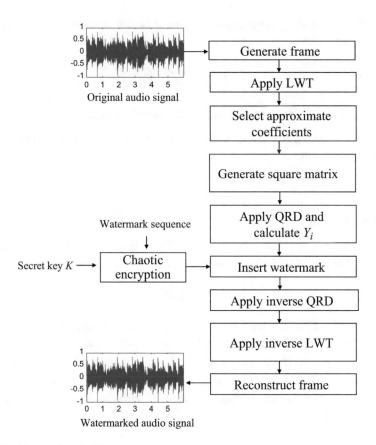

Fig. 3.1 Watermark embedding process

2. A two-level LWT is performed on each frame F_i. This operation produces three sets of coefficients D_1, D_2, and A_2, where D_1, D_2, and A_2 represent the detailed and approximate coefficients, respectively.
3. The approximate coefficients A_2 of each frame F_i are rearranged into an $N \times N$ square matrix H_i. This is done by dividing the coefficient set into N segments with N coefficients.
4. QRD is performed to decompose each matrix H_i into two matrices: upper triangular matrix R_i and unitary matrix Q_i. The QRD operation is represented as follows:

$$[Q_i, R_i] = qr(H_i) \tag{3.4}$$

5. In order to guarantee the robustness and transparency, the proposed method embeds watermark bits into the largest element of each matrix R_i using a quantization function. Let $Y_i = \text{round}\left(\dfrac{R_{i(\max)}}{D}\right)$, where D is a predefined quantization

coefficient and $R_{i(max)}$ is the largest element of each R_i. The embedding equation is given as follows:

$$Y_i' = \begin{cases} Y_i + C - (Y_i \bmod M), \text{if } u(i)=1 \\ Y_i + C - ((Y_i + C) \bmod M), \text{if } u(i)=0 \end{cases}$$ (3.5)

where $M = 2C$, C is an integer, mod is the modulo operation.

6. The modified largest element $R'_{i(max)}$ of each matrix R_i is calculated using the following equation:

$$R'_{i(max)} = Y_i' D$$ (3.6)

where R'_i is the upper triangular matrix.

7. Reinsert each modified largest element $R'_{i(max)}$ into each matrix R_i and inverse QRD is applied to obtain the modified matrix H_i', which is given by

$$H_i' = Q_i R_i'$$ (3.7)

Each matrix H_i' is then reshaped to create the modified approximate coefficients A_2' of each frame F_i by performing the inverse operation of step 3.

8. After substituting the modified approximate coefficients A_2' for A_2, a two-level inverse LWT is performed to obtain the watermarked audio frame F_i'.
9. Finally, all watermarked frames are concatenated to calculate the watermarked audio signal X'.

3.2.3   Watermark Detection Process

The proposed watermark detection process is shown in Fig. 3.2 and described as follows. It does not need the original audio signal to extract the watermark.

1. A two-level LWT is performed on each frame F_i^* of the attacked watermarked audio signal.
2. The approximate coefficients A_2^* of each frame F_i are rearranged into an $N \times N$ square matrix H_i^*.
3. QRD is performed on each H_i^* to get R_i^* and Q_i^*. The largest element $R_{i(max)}^*$ from each matrix R_i^* of the attacked watermarked audio frame is selected.
4. Calculate Y_i^* of each $R_{i(max)}^*$.
5. Encrypted watermark sequence is extracted as follows:

$$u^*(i) = \begin{cases} 1 \text{ if } (Y_i^* \bmod M)=1 \\ 0 \text{ otherwise} \end{cases}$$ (3.8)

Fig. 3.2 Watermark
detection process

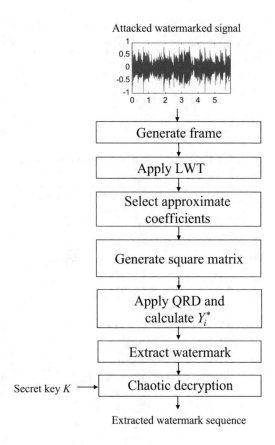

6. Perform chaotic decryption using the secret key K to find the binary watermark
   sequence with the following rule:

$$w^*(i) = z(i) \oplus u^*(i) \tag{3.9}$$

3.3   Simulation Results

In this section, several experiments were conducted on four different types of 16-bit
mono audio signals (Pop, Jazz, Classical, and Speech) sampled at 44.1 kHz to demonstrate the performance of the proposed method. Each audio file contains 262,144
samples (duration 5.94 s). Each audio signal is divided into frames of size 256
samples. Thus, the total number of frames is 1024. In each frame of audio signal, we
embed one bit watermark data. In this study, the selected values for $y(1)$, a, b, T, and
M are 0.4, 5.90, −0.39, 0.25, and 1, respectively. These parameters have been
selected in order to achieve a good trade-off among the conflicting requirements of
imperceptibility, robustness, and data payload.

Table 3.1 MOS, ODG and SNR results for different watermarked sounds

Types of Signal	MOS	ODG	SNR
Pop	4.9	−0.74	34.65
Jazz	5.0	−0.56	35.48
Classical	4.9	−0.79	34.93
Speech	4.8	−0.83	33.26
Average	*4.9*	*−0.73*	*34.58*

Table 3.2 SNR and MOS comparison between the proposed and several recent methods

Reference	Algorithm	SNR	MOS
[10]	DWT-based energy proportion	17.95	4.15
[9]	Wavelet-based entropy	22.46	4.38
[26]	EMD	24.12	–
[20]	Patchwork	24.95	4.67
[16]	FFT amplitude modification	25.70	–
[23]	DCT-SVD	32.53	4.71
Proposed	*LWT-QRD*	*34.58*	*4.90*

Table 3.3 NC and BER of the extracted watermark for different audio signals

Audio Signal	Attack Type	NC	BER (%)
Pop	No attack	1	0
	Noise addition	0.9919	0.9766
	Cropping	0.9926	0.8789
	Re-sampling	0.9894	1.2695
	Re-quantization	1	0
	MP3 compression	0.9822	2.1484
Jazz	No attack	1	0
	Noise addition	0.9943	0.6836
	Cropping	0.9992	0.0977
	Re-sampling	0.9918	0.9766
	Re-quantization	1	0
	MP3 compression	0.9748	3.0303
Classical	No attack	1	0
	Noise addition	0.9959	0.4883
	Cropping	0.9984	0.1953
	Re-sampling	0.9910	1.0742
	Re-quantization	1	0
	MP3 Compression	0.9721	3.3203
Speech	No attack	1	0
	Noise addition	0.9951	0.5865
	Cropping	0.9976	0.2930
	Re-sampling	0.9902	1.1719
	Re-quantization	1	0
	MP3 Compression	0.9737	3.1250

The performance of the proposed scheme was evaluated in terms of data payload, mean opinion score (MOS), objective difference grade (ODG), signal-to-noise ratio (SNR), bit error rate (BER), and normalized cross-correlation (NC) [22].

Perceptual quality of watermarked audio signal can be done by using subjective and objective evaluation tests. The subjective evaluation was carried out by blind listening test with ten subjects of different ages and the result is summarized in terms of MOS. The objective evaluation was conducted by calculating the ODG and SNR. The MOS, ODG, and SNR values of the different watermarked signals are shown in Table 3.1. We observed that the MOS, ODG, and SNR values range from 4.8 to 5.0, −0.83 to −0.56, and 33.26 to 35.48, respectively, indicating that the proposed watermarking scheme provides good imperceptible watermarked sound. The proposed method was compared with the several recent methods in terms of the SNR and MOS. This comparison is based on the reported results in [9–10, 16, 20, 23, 26] as shown in Table 3.2. From the comparison of results, it is seen that our proposed method outperforms the recent watermarking methods in terms of SNR and MOS, indicating the high transparency of the watermarked audio signals.

The proposed method utilizes chaotic encryption to enhance the security. The proposed watermark embedding and detection processes depend on the quantization parameter D and the secret key K. Thus, it is impossible to maliciously detect the watermark without them. Table 3.3 shows the robustness results of the proposed method in terms of NC and BER against several attacks for the audio signals "Pop," "Jazz," "Classical," and "Speech." The NC values are all above 0.97 and the BER values are below 4%. This clearly indicates a high robustness of the proposed

Table 3.4 A general comparison between the proposed and several recent methods sorted by data payload

Reference	Algorithm	Payload (bps)	Re-sampling BER (%)	Re-quantization BER (%)	MP3 compression BER (%)
Proposed	*LWT-QRD*	*172.39*	*1.27 (22.05 kHz)*	*0 (8 bits/sample)*	*3.32 (128 kbps)*
[9]	Wavelet-based entropy	172.28–86.14	9.1 (22.05 kHz)	–	6.7 (128 kbps)
[10]	DWT-based energy proportion	114.82	6.92 (22.05 kHz)	–	5.71 (80 kbps)
[26]	EMD	46.9–50.3	3 (22.05 kHz)	0 (8 bits/sample)	1 (32 kbps)
[24]	DWT-SVD	45.90	2 (22.05 kHz)	0 (8 bits/sample)	1 (32 kbps)
[23]	DCT-SVD	43	0 (22.05 kHz)	0 (8 bits/sample)	3 (32 kbps)
[28]	Histogram	3	0 (–)	0 (8 bits/sample)	15 (128 kbps)
[29]	DWT-based histogram	2	0 (16 kHz)	0 (8 bits/sample)	17.50 (64 kbps)
[15]	LWT	–	16.50 (36.750 kHz)	22.09 (8 bits/sample)	51.73 (128 kbps)

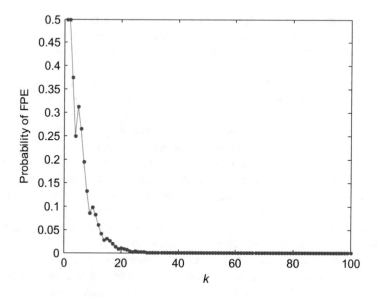

Fig. 3.3 Probability of FPE for various values of k

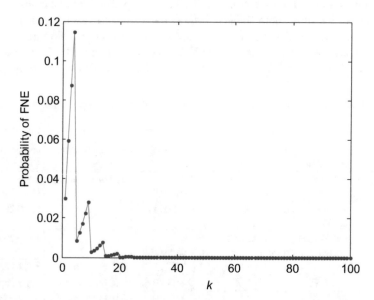

Fig. 3.4 Probability of FNE for various values of k

method against different attacks. This is because watermark bits are embedded into the largest element of the upper triangular matrix obtained from the low frequency LWT coefficients of each frame.

Table 3.4 shows a general comparison between the proposed and the several recent methods sorted by data payload, which is based on the reported result in the references [9–10, 15, 23–24, 26, 29]. Moreover, re-sampling, re-quantization, and MP3 compression are compared in Table 3.4. From this comparison, we observed that the proposed method provides higher data payload and lower BER values against several attacks compared with the recent watermarking methods.

Two types of errors may occur while searching the watermark sequence: (1) false positive error (FPE) and (2) false negative error (FNE). These errors are very harmful because they impair the credibility of watermarking method. The probability of FPE, P_{FPE}, and probability of FNE, P_{FNE}, can be calculated using Eqs. (2.11 and 2.12).

Figure 3.3 plots the P_{FPE} for $k \in (0,100]$. It is noted that P_{f_p} approaches 0 when k is larger than 30. Figure 3.4 plots the P_{FPE} for $k \in (0,100]$. It is noted that P_{f_n} approaches 0 when k is larger than 30.

3.4 Concluding Remarks

A blind audio watermarking method based on LWT and QRD has been presented in this chapter. The performance analysis of the proposed method indicates that it is highly robust against various attacks including noise addition, cropping, re-sampling, re-quantization, and MP3 compression. In addition, audio quality evaluation tests show high imperceptibility of the watermark in the audio signals. Moreover, it has high data payload and provides superior performance than the state-of-the-art audio watermarking methods. These results verify the effectiveness of the proposed watermarking method as a suitable candidate for audio copyright protection.

Chapter 4
Audio Watermarking Based on FWHT and LUD

4.1 Introduction

This chapter introduces an audio watermarking algorithm based on FWHT and LU decomposition (LUD). To the best of our knowledge, this is the first audio watermarking method based on FWHT, LUD, and quantization jointly. Initially, we preprocess the watermark data to enhance the security of the proposed algorithm. Then, the original audio is segmented into nonoverlapping frames and FWHT is applied to each frame. LUD is applied to the FWHT coefficients represented in a matrix form. Watermark data are embedded into the largest element of the upper triangular matrix obtained from the FWHT coefficients of each frame. Experimental results indicate that proposed algorithm is considerably robust and reliable against various attacks without degrading the quality of the watermarked audio. Moreover, it shows more excellent results than the state-of-the-art methods in terms of imperceptibility, robustness, and data payload. The main limitation of the existing audio watermarking techniques is the difficulty to obtain a favorable trade-off among imperceptibility, robustness, and data payload. To overcome this limitation, in this chapter, we propose a blind audio watermarking algorithm based on fast Walsh-Hadamard transform (FWHT) and LU decomposition (LUD). The main features of the proposed scheme are: (*i*) it utilizes the FWHT, LUD, and quantization jointly; (*ii*) it uses a tent map that contains the chaotic characteristic to enhance the confidentiality of the proposed algorithm; (*iii*) watermark is embedded into the largest element of the upper triangular matrix obtained from the FWHT coefficients of each frame by quantization; (*iv*) watermark extraction process is blind; and (*v*) it achieves a good trade-off among imperceptibility, robustness, and data payload. Experimental results indicate that the proposed watermarking algorithm shows high robustness against various attacks such as noise addition, cropping, re-sampling, re-quantization, and MP3 compression. Moreover, it outperforms state-of-the-art methods [9–10, 15–16, 20, 22–23, 26, 28] in terms of imperceptibility, robustness, and data payload.

4.2 Proposed Watermarking Method

Let $X = \{x(n), 1 \leq n \leq L\}$ be an original audio signal with L samples, $W = \{w(i), 1 \leq i \leq I\}$ be a binary watermark data to be embedded into the original audio signal.

4.2.1 Watermark Preprocessing

A tent map is utilized that contains the chaotic characteristics to encrypt the binary watermark data for enhancing the confidentiality of the proposed method. It can be defined as follows:

$$y(i+1) = \begin{cases} \frac{1}{\beta}y(i), & 0 \leq y(i) < \beta \\ \frac{1}{\beta-1}y(i) + \frac{1}{\beta-1}, & \beta \leq y(i) \leq 1 \end{cases} \tag{4.1}$$

where $y(1) \in (0,1)$ and β are real parameters (map's initial condition). Then $z(i)$ is calculated using the following rule:

$$z(i) = \begin{cases} 1 & \text{if } y(i) > T \\ 0 & \text{otherwise} \end{cases} \tag{4.2}$$

where T is a predefined threshold. Finally, $w(i)$ is encrypted by $z(i)$ with the following rule:

$$u(i) = z(i) \oplus w(i), 1 \leq i \leq I \tag{4.3}$$

where $\oplus$ is the exclusive-or (XOR) operation and $u(i)$ is the encrypted watermark data. After applying this chaotic encryption technique, the original watermark is encrypted and cannot be found by random search. In this study, the value of $y(1)$, a, and b are used as secret key K.

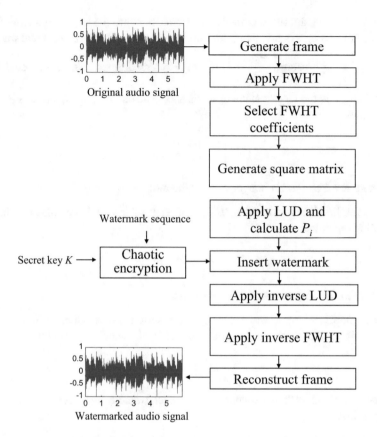

Fig. 4.1 Watermark embedding process

4.2.2   Watermark Embedding Process

The proposed watermark embedding process is shown in Fig. 4.1 and discussed as follows:

1. The original audio signal X is first segmented into nonoverlapping frames $F = \{F_1, F_2, F_3, \ldots\ldots, F_l\}$.
2. The FWHT is applied to each frame F_i to obtain the FWHT coefficients.
3. The FWHT coefficients of each frame F_i are rearranged into an $N \times N$ square matrix H_i. This is done by dividing the coefficient set into N segments with N coefficients.
4. LUD is performed to decompose each matrix H_i into three matrices: upper triangular matrix L_i, diagonal matrix D_i, and lower triangular matrix U_i. The LUD operation is represented as follows:

$$H_i = L_i D_i U_i \tag{4.4}$$

5. The proposed algorithm embeds watermark bits into the largest element of each matrix D_i using a quantization function to guarantee the robustness and transparency. Let $P_i = \text{round}\left(\dfrac{D_{i(\max)}}{Q}\right)$, where Q is a predefined quantization coefficient and $D_{i(\max)}$ is the largest element of each D_i. The embedding equation is given as follows:

$$P_i' = \begin{cases} P_i + C - (P_i \bmod R), & \text{if } u(i)=1 \\ P_i + C - \left((P_i + C)\bmod R\right), & \text{if } u(i)=0 \end{cases} \tag{4.5}$$

where $R = 2C$, C is an integer, mod is the modulo operation.

6. The modified largest element $D'_{i(\max)}$ of each matrix D_i is calculated using the following equation:

$$D'_{i(\max)} = P_i' Q \tag{4.6}$$

where D'_i is the modified diagonal matrix.

7. Each modified largest element $D'_{i(\max)}$ is reinserted into each matrix D_i and inverse LUD is applied to obtain the modified matrix H_i', which is given by

$$H_i' = U_i D_i' L_i \tag{4.7}$$

Each matrix H_i' is then reshaped to create each frame F_i by performing the inverse operation of step 3.

8. Each matrix H_i' is then reshaped to obtain the watermarked audio frame F_i'.
9. Finally, all watermarked frames are concatenated to calculate the watermarked audio signal X'.

4.2.3   Watermark Detection Process

The proposed blind watermark detection process is shown in Fig. 4.2 and discussed as follows:

1. The attacked watermarked audio signal X^* is segmented into nonoverlapping frames $F^* = \{F^*_1, F^*_2, F^*_3, \ldots, F^*_I\}$.
2. FWHT is applied to each frame and the FWHT coefficients of each frame F_i^* are rearranged into an $N \times N$ square matrix H_i^*.
3. LUD is performed on each H_i^* to get L_i^*, U_i^*, and D_i^*. The largest element $D^*_{i(\max)}$ from each matrix D_i^* of the attacked watermarked audio frame is selected.

4. Calculate P_i^* of each $D_{i(max)}^*$.
5. Encrypted watermark data are extracted as follows:

$$u^*(i) = \begin{cases} 1; & if\left(P_i^* \bmod R\right)=1 \\ 0; & otherwise \end{cases} \tag{4.8}$$

6. Perform chaotic decryption using the secret key K to find the binary watermark data with the following rule:

$$w^*(i) = z(i) \oplus u^*(i) \tag{4.9}$$

Fig. 4.2 Watermark detection process

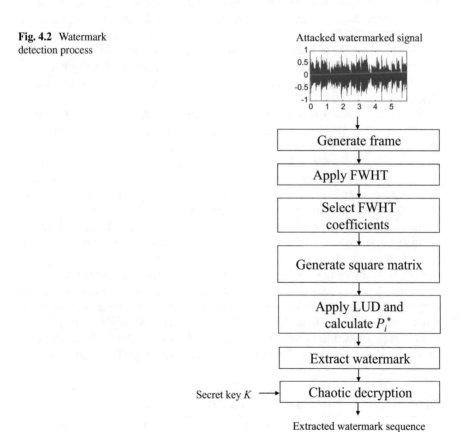

Attacked watermarked signal

Generate frame

Apply FWHT

Select FWHT coefficients

Generate square matrix

Apply LUD and calculate P_i^*

Extract watermark

Secret key K ⟶ Chaotic decryption

Extracted watermark sequence

4.3   Experimental Results and Discussion

In this section, several experiments were carried out on four different types of 16 bit mono audio signals (Pop, Jazz, Folk, and Speech) sampled at 44.1 kHz to demonstrate the performance of the proposed algorithm. Each audio file contains 262,144 samples (duration 5.94 s). Each audio signal is divided into frames of size 256 samples. Thus, the total number of frames is 1024. In each frame of audio signal, we embed one bit watermark data. Here, the selected values for $y(1)$, β, T, B, θ, and Q are 0.6, 0.3, 0.5, 2, 45°, and 0.4, respectively. These parameters were selected in order to achieve a good trade-off among the imperceptibility, robustness, and data payload.

We have evaluated the performance of the proposed algorithm in terms of data payload, mean opinion score (MOS), correct detection, objective difference grade (ODG), signal-to-noise ratio (SNR), bit error rate (BER), and normalized cross-correlation (NC) [22].

Perceptual quality of watermarked audio signal can be evaluated using subjective and objective tests. The subjective listening test was carried out blind by ten listeners of different ages and the result is summarized in terms of MOS and correct detection.

The objective test was conducted by calculating the ODG and SNR. The MOS, correct detection, ODG, and SNR values of the different watermarked signals are shown in Table 4.1. It is seen that the MOS, correct detection, ODG, and SNR values range from 4.8 to 4.9, 46% to 58%, −0.72 to −0.53, and 35.47 to 36.65, respectively, indicating that the proposed watermarking algorithm provides good imperceptible watermarked sound. Table 4.2 shows a comparison between the proposed algorithm and the several recent methods in terms of SNR and MOS that are based on the reported results in the references [9–10, 16, 20, 23, 26]. From this comparison, it is observed that the proposed algorithm shows better result than the recent watermarking methods in terms of SNR and MOS. In other words, subjective and objective tests prove the high transparency of the proposed algorithm.

Various signal processing attacks were applied to assess the robustness of the proposed algorithm. Table 4.3 shows the NC and BER results of the proposed algorithm against various attacks for different audio signals. We observed that the NC values range from 0 to 0.9862 and the BER values range from 0% to 2%. This clearly indicates a good performance of the proposed algorithm against various attacks.

Table 4.1 Subjective and objective evaluation of different watermarked sounds

Types of signal	Subjective evaluation		Objective evaluation	
	MOS	Correct detection	SNR	ODG
Pop	4.90	58%	36.65	−0.53
Jazz	4.90	54%	35.47	−0.58
Classical	4.90	48%	36.39	−0.62
Speech	4.80	46%	35.78	−0.72
Average	*4.88*	*51.5%*	*36.07*	*−0.61*

Table 4.2 SNR and MOS comparison between the proposed and several recent methods

Reference	Algorithm	SNR	MOS
[10]	DWT-based energy proportion	17.95	4.15
[9]	Wavelet-based entropy	22.46	4.38
[26]	EMD	24.12	–
[20]	Patchwork	24.95	4.67
[16]	FFT amplitude modification	25.70	–
[23]	DCT-SVD	32.53	4.71
Proposed	*FWHT-LUD*	*36.07*	*4.88*

Table 4.3 NC and BER of the extracted watermark for different audio signals

Audio signal	Attack type	NC	BER (%)
Pop	No attack	1	0
	Noise addition	0.9964	0.5254
	Cropping	0.9983	0.2320
	Re-sampling	1	0
	Re-quantization	1	0
	MP3 compression	0.9894	1.2695
Jazz	No attack	1	0
	Noise addition	1	0
	Cropping	0.9926	0.8789
	Re-sampling	1	0
	Re-quantization	1	0
	MP3 compression	0.9918	0.9766
Folk	No attack	1	0
	Noise addition	0.9926	0.8789
	Cropping	1	0
	Re-sampling	1	0
	Re-quantization	1	0
	MP3 Compression	0.9862	1.7465
Speech	No attack	1	0
	Noise addition	0.9951	0.5865
	Cropping	0.9976	0.2930
	Re-sampling	1	0
	Re-quantization	1	0
	MP3 Compression	0.9902	1.1719

A general comparison between the proposed algorithm and the several recent methods sorted by data payload, which is based on the reported result in the references [9, 15, 22, 23, 26, 28], is presented in Table 4.4. Moreover, Table 4.4 shows a comparison for re-sampling, re-quantization, and MP3 compression. From this comparison, we can conclude that the proposed algorithm has higher data payload and lower BER values against various attacks than the state-of-the-art methods. This is because watermark bits are embedded into the largest element of the diagonal matrix obtained from the FWHT coefficients of each frame.

Table 4.4 A general comparison between the proposed algorithm and several recent methods sorted by data payload

Reference	Algorithm	Payload (bps)	Re-sampling BER (%)	Re-quantization BER (%)	MP3 compression BER (%)
Proposed	*FWHT-LUD*	*172.39*	*0 (22.05 kHz)*	*0 (8 bits/sample)*	*1.75 (128 kbps)*
[22]	DCT-SVD-LPT	172.39	1.56 (22.05 kHz)	0 (8 bits/sample)	3.91 (128 kbps)
[9]	Wavelet-based entropy	86.14~172.28	9.1 (22.05 kHz)	–	6.7 (128 kbps)
[26]	EMD	46.9~50.3	3 (22.05 kHz)	0 (8 bits/sample)	1 (32 kbps)
[23]	DCT-SVD	43	0 (22.05 kHz)	0 (8 bits/sample)	3 (32 kbps)
[28]	Histogram	3	0 (–)	0 (8 bits/sample)	15 (128 kbps)
[15]	LWT	–	16.50 (36.750 kHz)	22.09 (8 bits/sample)	51.73 (128 kbps)

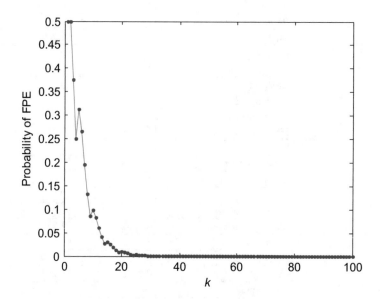

Fig. 4.3 Probability of FPE for various values of k

The false positive error (FPE) and false negative error (FNE) are very harmful because they impair the credibility of watermarking method. The probability of FPE, P_{FPE}, and probability of FNE, P_{FNE}, can be calculated using Eqs. (2.11 and 2.12).

Figure 4.3 plots the P_{FPE} for $k \in (0,100]$. It is noted that P_{f_p} approaches 0 when k is larger than 30. Figure 4.4 plots the P_{FPE} for $k \in (0,100]$. It is noted that P_{f_n} approaches 0 when k is larger than 30.

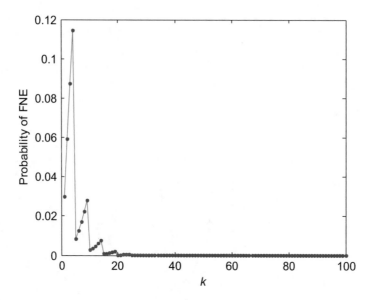

Fig. 4.4 Probability of FNE for various values of k

4.4 Concluding Remarks

In this chapter, we introduced a blind audio watermarking algorithm based on FWHT and LUD. Experimental results demonstrate that the embedding data are robust against various attacks such as noise addition, cropping, re-sampling, re-quantization, and MP3 compression. In addition, subjective and objective tests show high imperceptibility of the watermarked signals. Moreover, it has high data payload and provides superior performance than the state-of-the-art audio watermarking methods. These results verify that the proposed algorithm can be effectively utilized for audio copyright protection. In future, we will extend our research to image and video watermarking.

Chapter 5
Audio Watermarking Based on LWT and SD

5.1 Introduction

Chapter 5 proposes an audio watermarking method based on LWT and Schur decomposition (SD). To the best of our knowledge, this is the first audio watermarking method based on LWT, SD, and quantization jointly. Initially, the watermark data are preprocessed to enhance the confidentiality of the proposed method. Then, the original audio is segmented into nonoverlapping frames and LWT is applied to each frame. SD is applied to the selected low-frequency LWT coefficients represented in a matrix form. Watermark data are embedded into the largest element of the upper triangular matrix obtained from the selected LWT coefficients of each frame. Experimental results confirm that the embedded data are highly robust against various attacks. Moreover, it shows superior performance than the state-of-the-art watermarking methods reported recently. In this chapter, an audio watermarking method in lifting wavelet transform (LWT) domain based on Schur decomposition (SD) is introduced. The main features of the proposed method are: (*i*) it utilizes the LWT and SD jointly, (*ii*) it uses Gaussian map, containing the chaotic characteristic to enhance the confidentiality of the proposed scheme, (*iii*) watermark extraction process is blind, (*iv*) subjective and objective evaluations reveal that the proposed scheme maintains high audio quality, and (*v*) it achieves a good trade-off among imperceptibility, robustness, and data payload. Experimental results indicate that the proposed watermarking scheme is highly robust against various attacks such as noise addition, cropping, re-sampling, re-quantization, and MP3 compression. Moreover, it outperforms state-of-the-art methods [9–10, 14–16, 20, 23–24, 26] in terms of imperceptibility, robustness, and data payload. The data payload of the proposed scheme is 172.39 bps, which is relatively higher than that of the state-of-the-art methods.

5.2   Proposed Watermarking Method

In this section, we give an overview of our basic watermarking scheme, which consists of watermark preprocessing, watermark embedding and watermark detection processes.

Let $X = \{x(n), 1 \leq n \leq L\}$ be an original audio signal with L samples, and $W = \{w(i), 1 \leq i \leq I\}$ be a binary watermark sequence to be embedded into the original audio signal.

5.2.1   Watermark Preprocessing

To enhance the security of the proposed method, watermark sequence should be preprocessed first. The proposed method utilizes a Gaussian map, containing the chaotic characteristics to encrypt the binary watermark sequence for enhancing the confidentiality of the proposed method. It can be defined as follows:

$$y(i+1) = \exp\left(-a\left(y(i)\right)^2\right) + b \qquad (5.1)$$

where $y(1) \in (0,1)$, a and b are real parameters (map's initial condition). Then $y(i)$ is mapped using the following rule:

$$z(i) = \begin{cases} 1 & \text{if } y(i) > T \\ 0 & \text{otherwise} \end{cases} \qquad (5.2)$$

where T is a predefined threshold. Finally, $w(i)$ is encrypted by $z(i)$ with the following rule:

$$u(i) = z(i) \oplus w(i), 1 \leq i \leq I \qquad (5.3)$$

where $\oplus$ is the exclusive-or (XOR) operation and $u(i)$ is the encrypted watermark sequence. After applying this chaotic encryption technique, the original watermark is encrypted and cannot be found by random search. In this study, the value of $y(1)$, a, and b are used as secret key K.

5.2.2   Watermark Embedding Process

The proposed watermark embedding process is shown in Fig. 5.1 and described as follows:

1. The original audio signal X is first segmented into nonoverlapping frames $F = \{F_1, F_2, F_3, \ldots, F_I\}$.

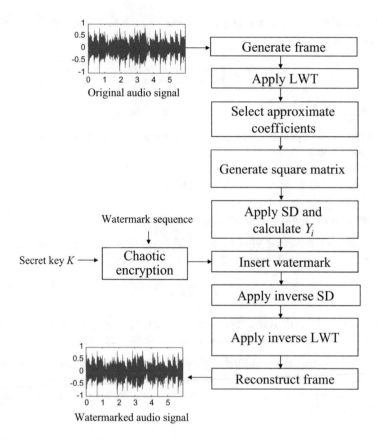

Fig. 5.1 Watermark embedding process

2. A two-level LWT is performed on each frame F_i. This operation produces three sets of coefficients D_1, D_2, and A_2, where D_1, D_2, and A_2 represent the detailed and approximate coefficients, respectively.

3. The approximate coefficients A_2 of each frame F_i are rearranged into an $N \times N$ square matrix H_i. This is done by dividing the coefficient set into N segments with N coefficients.

4. SD is performed to decompose each matrix H_i into two matrices: upper triangular matrix S_i and unitary matrix U_i. The SD operation is represented as follows:

$$H_i = U_i S_i U_i^{-1} \tag{5.4}$$

5. In order to guarantee the robustness and transparency, the proposed method embeds watermark bits into the largest element of each matrix S_i using a quantization function. Let $Y_i = \mathrm{round}\left(\dfrac{S_{i(\max)}}{D}\right)$, where D is a predefined quantization

coefficient and $S_{i(max)}$ is the largest element of each S_i. The embedding equation is given as follows:

$$Y_i = \begin{cases} Y_i + C - (Y_i \bmod M), & \text{if } u(i)=1 \\ Y_i + C - ((Y_i + C) \bmod M), & \text{if } u(i)=0 \end{cases} \tag{5.5}$$

where $M = 2C$, C is an integer, mod is the modulo operation.

6. The modified largest element $S'_{i(max)}$ of each matrix S_i is calculated using the following equation:

$$S_{i(max)} = Y_i D \tag{5.6}$$

where S'_i is the modified upper triangular matrix.

7. Reinsert each modified largest element $S'_{i(max)}$ into each matrix S_i and inverse SD is applied to obtain the modified matrix H_i', which is given by

$$H_i = U_i S_i U_i^{-1} \tag{5.7}$$

Each matrix H_i' is then reshaped to create the modified approximate coefficients A'_2 of each frame F_i by performing the inverse operation of step 3.

8. After substituting the modified approximate coefficients A'_2 for A_2, a two-level inverse LWT is performed to obtain the watermarked audio frame F_i'.
9. Finally, all watermarked frames are concatenated to calculate the watermarked audio signal X'.

5.2.3   Watermark Detection Process

The proposed blind watermark detection process is shown in Fig. 5.2 and described as follows:

1. A two-level LWT is performed on each frame F_i^* of the attacked watermarked audio signal.
2. The approximate coefficients A_2^* of each frame F_i are rearranged into an $N \times N$ square matrix H_i^*.

Table 5.1 MOS, ODG, and SNR results for different watermarked sounds

Types of signal	MOS	ODG	SNR
Pop	4.80	−0.77	33.75
Jazz	4.90	−0.59	34.84
Classical	4.90	−0.72	33.39
Speech	4.80	−0.85	34.67
Average	4.85	−0.73	34.16

Fig. 5.2 Watermark
detection process

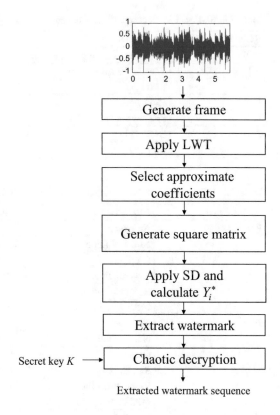

Extracted watermark sequence

3. SD is performed on each H_i^* to get S_i^* and U_i^*. The largest element $S^*_{i(\max)}$ from each matrix S_i^* of the attacked watermarked audio frame is selected.
4. Calculate Y_i^* of each $S^*_{i(\max)}$.

Encrypted watermark sequence is extracted as follows:

$$u^* (i) = \begin{cases} 1 & \text{if } (Y_i^* \bmod M)=1 \\ 0 & \text{otherwise} \end{cases} \tag{5.8}$$

5. Perform chaotic decryption using the secret key K to find the binary watermark sequence with the following rule:

$$w^* (i) = z(i) \oplus u^* (i) \tag{5.9}$$

Table 5.2 SNR and MOS comparison between the proposed and several recent methods

Reference	Algorithm	SNR	MOS
[10]	DWT-based energy proportion	17.95	4.15
[9]	Wavelet-based entropy	22.46	4.38
[26]	EMD	24.12	–
[20]	Patchwork	24.95	4.67
[16]	FFT amplitude modification	25.70	–
[23]	DCT-SVD	32.53	4.71
Proposed	*LWT-SD*	*34.16*	*4.85*

Table 5.3 NC and BER of the extracted watermark for different audio signals

Audio signal	Attack type	NC	BER (%)
Pop	No attack	1	0
	Noise addition	0.9951	0.5859
	Cropping	0.9976	0.2930
	Re-sampling	0.9894	1.2695
	Re-quantization	1	0
	MP3 compression	0.9721	3.3203
Jazz	No attack	1	0
	Noise addition	0.9919	0.9766
	Cropping	0.9926	0.8789
	Re-sampling	0.9894	1.2695
	Re-quantization	1	0
	MP3 compression	0.9822	2.1484
Classical	No attack	1	0
	Noise addition	0.9926	0.8789
	Cropping	0.9984	0.1953
	Re-sampling	0.9918	0.9766
	Re-quantization	1	0
	MP3 Compression	0.9812	2.2461
Speech	No attack	1	0
	Noise addition	0.9951	0.5865
	Cropping	0.9976	0.2930
	Re-sampling	0.9902	1.1719
	Re-quantization	1	0
	MP3 Compression	0.9737	3.1250

5.3   Experimental Results and Discussion

In this section, we conducted several experiments on four different types of 16-bit mono audio signals (Pop, Jazz, Classical, and Speech) sampled at 44.1 kHz to demonstrate the performance of the proposed method. Each audio file contains 262,144 samples (duration 5.94 s). Each audio signal is divided into frames of size 256 samples. Thus, the total number of frames is 1024. In each frame of audio signal, we

Table 5.4 A general comparison between the proposed scheme and the several recent methods sorted by data payload

Reference	Algorithm	Payload (bps)	Re-sampling BER (%)	Re-quantization BER (%)	MP3 compression BER (%)
Proposed	*LWT-SD*	*172.39*	*1.27 (22.05 kHz)*	*0 (8 bits/sample)*	*3.32 (128 kbps)*
[9]	Wavelet-based entropy	172.28–86.14	9.1 (22.05 kHz)	–	6.7 (128 kbps)
[14]	Chaos based DFRST	86	0 (22.05 kHz)	–	3.47 (48 kbps)
[26]	EMD	46.9–50.3	3 (22.05 kHz)	0 (8 bits/sample)	1 (32 kbps)
[23]	DCT-SVD	43	0 (22.05 kHz)	0 (8 bits/sample)	3 (32 kbps)
[24]	Histogram	3	0 (–)	0 (8 bits/sample)	15 (128 kbps)
[15]	LWT	–	16.50 (36.750 kHz)	22.09 (8 bits/sample)	51.73 (128 kbps)

embed one bit watermark data. In this study, the selected value for $y(1)$, a, b, T, and M are 0.4, 5.90, −0.39, and 0.25, respectively. These parameters were selected in order to achieve a good trade-off among the conflicting requirements of imperceptibility, robustness, and data payload.

The performance of the proposed scheme was evaluated in terms of data payload, mean opinion score (MOS), objective difference grade (ODG), signal-to-noise ratio (SNR), bit error rate (BER), and normalized cross-correlation (NC) [22].

Perceptual quality of watermarked audio signal can be evaluated by using subjective and objective evaluation tests. The subjective evaluation was carried out by blind listening test with ten subjects of different ages and the result is summarized in terms of MOS. The objective evaluation was conducted by calculating the ODG and SNR. The MOS, ODG, and SNR values of the different watermarked signals are shown in Table 5.1. We observed that the MOS, ODG, and SNR values range from 4.8 to 4.9, −0.85 to −0.59, and 33.39 to 34.67, respectively, indicating that the proposed watermarking scheme provides good imperceptible watermarked sound. Table 5.2 shows a comparison between the proposed scheme and the several recent methods in terms of SNR and MOS that are based on the reported results in the references [9–10, 16, 20, 23, 26]. From these results, we observed that our proposed scheme provides better result than the recent watermarking methods in terms of SNR and MOS. In other word, subjective and objective evaluations prove a high transparency of the proposed method.

Various signal processing attacks were performed to assess the robustness of the proposed scheme. Table 5.3 shows the robustness results of the proposed scheme in terms of NC and BER against several attacks for the audio signals "Pop," "Jazz," "Classical," and "Speech," respectively. The NC values are all above 0.97 and the BER values are below 4%. This clearly indicates a good performance of the proposed method against different attacks, demonstrating the high robustness of our proposed method against different attacks.

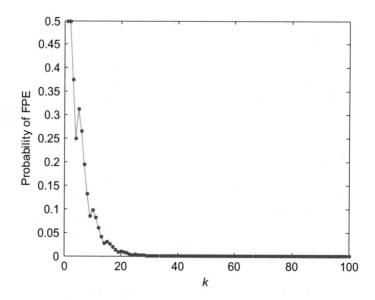

Fig. 5.3 Probability of FPE for various values of k

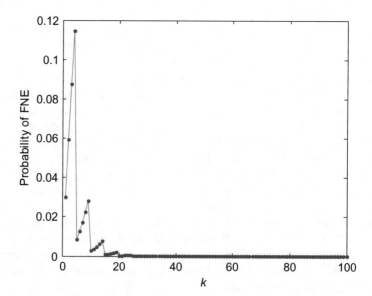

Fig. 5.4 Probability of FNE for various values of k

Table 5.4 shows a general comparison between the proposed and the several recent methods sorted by data payload, which is based on the reported result in the references [9, 14–15, 23, 24, 26]. Moreover, re-sampling, re-quantization, and MP3 compression are compared in Table 5.4. From this comparison, we observed that the proposed method provides higher data payload and lower BER values against several attacks compared with the recent watermarking methods. This is because watermark bits are embedded into the largest element of the upper triangular matrix obtained from the low-frequency LWT coefficients of each frame.

The false positive error (FPE) and false negative error (FNE) are very harmful because they impair the credibility of watermarking method. The probability of FPE, P_{FPE}, and probability of FNE, P_{FNE}, can be calculated using Eqs. (2.11 and 2.12).

Figure 5.3 plots the P_{FPE} for $k \in (0,100]$. It is noted that P_{f_p} approaches 0 when k is larger than 30. Figure 5.4 plots the P_{FPE} for $k \in (0,100]$. It is noted that P_{f_n} approaches 0 when k is larger than 30.

5.4 Concluding Remarks

A blind audio watermarking method based on LWT and SD was presented in this chapter. Simulation results indicate that the proposed method is highly robust against various attacks including noise addition, cropping, re-sampling, re-quantization, and MP3 compression. In addition, audio quality evaluation tests show high imperceptibility of the watermark in the audio signals. Moreover, it has high data payload and provides better result than the state-of-the-art audio watermarking methods. These results verify that the proposed method can be a suitable candidate for audio copyright protection.

Chapter 6
Conclusions and Future Work

6.1 Conclusion

This chapter concludes this book with a brief summary of our research work. Future research work is also discussed in this chapter.

Digital watermarking is a process of embedding a secret signal called the watermark within the original signal to show authenticity and ownership. It has been utilized effectively to provide solutions for ownership protection, copyright protection, content authentication, speech quality evaluation, secret communication. Depending on the watermark application and purpose, two important issues need to be addressed for digital audio watermarking. One is to provide trustworthy evidence to protect rightful ownership and the other is to achieve an appropriate trade-off among imperceptibility, robustness, and data payload. The audio watermarking methods proposed in this book are summarized as follows:

- A blind LWT-based watermarking method using FWHT and SVD has been presented for audio copyright protection. Watermark data are embedded into the largest singular value of the FWHT coefficients obtained from the low-frequency LWT coefficients of each frame using a quantization function. This method provides high imperceptible watermarked sounds as well as good robustness against various attacks. The comparison analysis indicates that the proposed scheme outperforms the state-of-the-art watermarking methods reported recently.
- A blind audio watermarking method based on LWT and QRD has been introduced for audio copyright protection. Watermark information is embedded into the largest element of the upper triangular matrix obtained from the low-frequency LWT coefficients of each frame. A blind watermark detection technique is introduced to identify the embedded watermark under various attacks. This method has high data payload and provides good imperceptible watermarked

© The Author(s), under exclusive license to Springer Nature Switzerland AG 2019
P. K. Dhar, T. Shimamura, *Advances in Audio Watermarking Based on Matrix Decomposition*, SpringerBriefs in Speech Technology,
https://doi.org/10.1007/978-3-030-15726-5_6

sounds. Moreover, it shows better result than the state-of-the-art methods in terms of imperceptibility and robustness.

- An audio watermarking algorithm based on FWHT and LUD has been proposed. To the best of our knowledge, this is the first audio watermarking method based on FWHT, LUD, and quantization jointly. Watermark data are embedded into the largest element of the upper triangular matrix obtained from the FWHT coefficients of each frame. The proposed algorithm is considerably robust and reliable against various attacks without degrading the quality of the watermarked audio. Moreover, it shows more excellent results than the state-of-the-art methods in terms of imperceptibility, robustness, and data payload.

- An audio watermarking method based on LWT and SD has been presented. To the best of our knowledge, this is the first audio watermarking method based on LWT, SD, and quantization jointly. Watermark data are embedded into the largest element of the upper triangular matrix obtained from the selected LWT coefficients of each frame. Experimental results confirm that the embedded data are highly robust against various attacks. Moreover, it shows superior performance than the state-of-the-art watermarking methods reported recently.

6.2 Future Work

There are several directions for future research on the proposed methods introduced in this book. In the future work, synchronization code and error correcting codes will be incorporated to improve the robustness of the proposed methods. In addition, the adaptive selection of quantization parameter Q might further improve the performance of the proposed methods. Some modern attacks such as channel fading, jitter, and packet drop will be considered, because these attacks are particularly relevant in various networks such as GSM (Global System for Mobile communications) and CDMA (Code Division Multiple Access). Moreover, psychoacoustic model can be adopted to improve the imperceptibility of the proposed methods. Furthermore, computational complexity of the proposed methods will be carried out.

References

1. I.J. Cox, M.L. Miller, J. Bloom, J. Fridrich, T. Kalker, *Digital Watermarking and Steganography*, (The Morgan Kaufmann Series in Multimedia Information Systems, Elsevier, USA, 2008)
2. I.J. Cox, M.L. Miller, The first 50 years of electronic watermarking. J. Applied Signal Process. **56**(2), 225–230 (2002)
3. S. Katzenbeisser, F.A.P. Petitcolas, *Information Hiding Techniques for Steganography and Digital Watermarking* (Artech House Norwood Mass, Boston, 2000)
4. N. Cvejic, T. Seppanen, *Digital Audio Watermarking Techniques and Technologies* (Information Science Reference, Hershey, 2007)
5. P.K. Dhar, T. Shimamura, *Advances in Audio Watermarking Based on Singular Value Decomposition* (Springer, Cham, 2015)
6. P. K. Dhar, Studies on digital audio watermarking using matrix decomposition, Ph.D thesis, Saitama University, (2014)
7. W.N. Lie, L.C. Chang, Robust high quality time domain audio watermarking based on low frequency amplitude modification. IEEE Trans. Multimedia **8**(1), 46–59 (2006)
8. P. Bassia, I. Pitas, N. Nikolaidis, Robust audio watermarking in the time domain. IEEE Trans. Multimedia **3**(2), 232–241 (2001)
9. S.T. Chen, H.N. Huang, C.J. Chen, K.K. Tseng, S.Y. Tu, Adaptive audio watermarking via the optimization point of view on the wavelet-based entropy. Digital Signal Process. **23**(3), 971–980 (2013)
10. S.T. Chen, H.N. Huang, C.J. Chen, G.D. Wu, Energy-proportion based scheme for audio watermarking. IET Signal Process. **4**(5), 576–587 (2010)
11. S.T. Chen, G.D. Wu, H.N. Huang, Wavelet-domain audio watermarking scheme using optimisation-based quantization. IET Signal Process. **4**(6), 720–727 (2010)
12. M.K. Dutta, P. Gupta, V.K. Pathak, A perceptible watermarking algorithm for audio signals. Multimed. Tools Appl. **73**, 1–23 (2012)
13. H.H. Tsai, J.S. Cheng, P.T. Yu, Audio watermarking based on HAS and neural networks in DCT domain. EURASIP J. Adv. Signal Process. **3**, 252–263 (2003)
14. M. Fan, H. Wang, Chaos-based discrete fractional sine transform domain audio watermarking scheme. Comp. Elect. Eng. **35**(3), 506–516 (2009)
15. E. Erçelebi, L. Batakçı, Audio watermarking scheme based on embedding strategy in low frequency components with a binary image. Digital Signal Process. **19**(2), 265–277 (2009)
16. D. Megías, J. Serra-Ruiz, M. Fallahpour, Efficient self–synchronized blind audio watermarking system based on time domain and FFT amplitude modification. Signal Process. **90**(12), 3078–3092 (2010)

17. M. Fallaphour, D. Megias, Robust high-capacity audio watermarking based on FFT amplitude modification. IEICE Trans. Inf. Sys. **E93-D**(1), 87–93 (2010)
18. I. Natgunanathan, Y. Xiang, Y. Rong, W. Zhou, S. Guo, Robust patchwork-based embedding and decoding scheme for digital audio watermarking. IEEE Trans. Audio Speech Lang. Process. **20**(8), 2232–2239 (2012)
19. I.K. Yeo, H.J. Kim, Modified patchwork algorithm: A novel audio watermarking scheme. IEEE Trans. Speech Audio Process. **11**(4), 381–386 (2003)
20. H. Kang, K. Yamaguchi, B. Kurkoski, K. Yamaguchi, K. Kobayashi, Full-index-embedding patchwork algorithm for audio watermarking. IEICE Trans. Inf. & Syst. **E91-D**(11), 2731–2734 (2008)
21. P.K. Dhar, T. Shimamura, Audio watermarking in transform domain based on singular value decomposition and Cartesian-polar transformation. Int. J. Speech Technol. **17**, 133–144 (2014)
22. P.K. Dhar, T. Shimamura, Blind SVD-based audio watermarking using entropy and log-polar transformation. J. Information Security and Application **20**, 74–83 (2015)
23. B.Y. Lei, I.Y. Soon, Z. Li, Blind and robust audio watermarking scheme based on SVD-DCT. Signal Process. **91**, 1973–1984 (2011)
24. V.K. Bhat, I. Sengupta, A. Das, An adaptive audio watermarking based on the singular value decomposition in the wavelet domain. Digital Signal Process. **20**(6), 1547–1558 (2010)
25. W. Al-Nuaimy, M.A.M. El-Bendary, A. Shafik, F. Shawki, A.E. Abou-El-Azm, N.A. El-Fishawy, S.M. Elhalafawy, S.M. Diab, B.M. Sallam, F.E.A. El-Samie, H.B. Kazemian, An SVD audio watermarking approach using chotic encrypted images. Digital Signal Process. **21**(6), 764–779 (2011)
26. K. Khaldi, A.O. Boudraa, Audio watermarking via EMD. IEEE Trans. Audio Speech Lang. Process. **21**(3), 675–680 (2013)
27. Y. Xiang, I. Natgunanathan, D. Peng, W. Zhou, S. Yu, A dual-channel time-spread echo method for audio watermarking. IEEE Trans. Inf. Forensics Security **7**(2), 383–392 (2012)
28. S. Xiang, J. Huang, Histogram based audio watermarking against time scale modification and cropping attacks. IEEE Trans. Multimedia **9**(7), 1357–1372 (2007)
29. S. Xiang, H.J. Kim, J. Huang, Audio watermarking robust against time scale modification and MP3 compression. Signal Process. **88**(10), 2372–2387 (2008)
30. P.K. Dhar and T. Shimamura, A blind LWT-based audio watermarking using fast Walsh-Hadamard transform and singular value decomposition, in *Proceedings of IEEE International Symposium on Circuits and Systems (ISCAS-2014)*, Melbourne, Australia, 1–5 June 2014, pp, 125–128
31. P.K. Dhar, A blind audio watermarking method based on lifting wavelet transform and QR decomposition, in *Proceedings of the 8th International Conference on Electrical and Computer Engineering (ICECE-2014)*, Dhaka, Bangladesh, 20–22 Dec 2014, pp. 136–139
32. T. Thiede, W.C. Treurniet, R. Bitto, C. Schmidmer, T. Sporer, J.G. Beerens, C. Colomes, M. Keyhl, G. Stoll, K. Brandenburg, B. Feiten, PEAQ - the ITU standard for objective measurement of perceived audio quality. J. Audio Eng. Soc. **48**(1/2), 3–29 (2000)

Printed in the United States
By Bookmasters